D0903538

PIONEERS OF SCIENCE

KARL BENZ

Brian Williams

The Bookwright Press
New York · 1991

Pioneers of Science

Archimedes
Alexander Graham Bell
Karl Benz
Marie Curie
Thomas Edison
Albert Einstein
Michael Faraday
Galileo
Guglielmo Marconi
Isaac Newton
Louis Pasteur
Leonardo da Vinci

First published in the
United States in 1991 by
The Bookwright Press
387 Park Avenue South
New York, NY 10016

First published in 1991 by
Wayland (Publishers) Ltd
61 Western Road, Hove
East Sussex BN3 1JD, England

Library of Congress Cataloging–in–Publication Data
Williams, Brian.
Karl Benz / Brian Williams.
p. cm. – (Pioneers of science)
Includes bibliographical references and index.
Summary: A biography of the German engineer and inventor who pioneered in building motor-driven vehicles.
ISBN 0–531–18404–8
1. Benz, Karl, 1844–1929 – Juvenile literature.
2. Automobile engineers – Germany – Biography – Juvenile literature. [1. Benz, Karl, 1844–1929.
2. Automobile engineers. 3. Engineers.
4. Inventors.] I. Title. II. Series.
TL140.B4W55 1991
629.2'092–dc20
[B]
[92] 90–21744
CIP
AC

Typeset by Kalligraphic Design Ltd, Horley
Printed in Italy by Rotolito Lombardo S.p.A.

Contents

1 Early Life

No invention has changed daily life more than the automobile. Highways dominate the environment, and the auto industry employs millions of people worldwide. It is difficult for someone living in North America, Europe or Japan to imagine a world without any cars.

When Karl Benz was born in 1844, the automobile was unknown. The railroad age was scarcely fifteen years old. The fastest that most people had ever traveled was on the back of a horse or in a horse-drawn carriage. Yet the steam and smoke rising from the new factories and workshops of Britain, France and Germany heralded a new age. An industrial revolution was under way, and new technologies were about to transform people's lives.

A busy London street in the early 1800s. In cities, horse-drawn vehicles, large and small, were causing traffic jams long before the automobile appeared.

Karl Benz was born on November 25, 1844, at Karlsruhe in Germany. At that time, Germany was one of Europe's leaders in the Industrial Revolution. Karl's father, Johann, was a locomotive driver on the newly opened railroad. Johann died of pneumonia when the boy was only two years old, following an accident on the railroad.

Karl's mother, Josephine, faced a hard struggle to bring up her son on a small widow's pension. She gave up the family home at Muhlberg, outside Karlsruhe, and moved into the town itself. There she earned extra money as a cook. Josephine Benz was determined that her son should have a good education; her ambition was for him to become a civil servant.

Karl went to the local lyceum (high school). Unlike many schools of the time, its curriculum included a generous amount of science. Paying for Karl's schooling cost Josephine Benz one-sixth of her year's pension. His favorite subject was physics, and somehow she scraped together enough money to buy some laboratory equipment so that Karl could work on scientific experiments in the attic at home.

A modern Mercedes car, the descendant of Karl Benz's three-wheeler of 1885.

From school, Karl went on to the Fridericiana Technical College in Karlsruhe. It was founded in 1825 and was the first of its kind in Germany. Anyone could study there, but the fees were a problem for the less well-off. To boost her income, Josephine Benz took in lodgers, and Karl repaired watches in his spare time.

Karl had an outstanding science teacher named Ferdinand Redtenbacher, who introduced the eager teenager to the new "explosion engine." It offered an alternative to the steam engine – one with pistons and cylinders powered not by steam but by exploding gas. The first working gas engine was built by a Frenchman, Etienne Lenoir, in 1860. It resembled a steam engine, but burned coal gas from the municipal supply. A mixture of gas and air was drawn into the cylinder and ignited. The gas exploded, expanded and pushed the cylinder down inside the chamber. It was not very powerful, but it was quiet, fairly reliable and smaller and cleaner than a steam engine. It was useful in small workshops, such as those of printers and shoemakers.

Below *Etienne Lenoir's gas engine of 1860 worked much the same way a steam engine does. Using a belt drive, the engine could power small machinery.*

Opposite Trevithick's steam passenger coach of 1803. This was a promising experiment, but it was slow, hard to steer, and potentially dangerous.

In 1861, a Lenoir gas engine was installed in the Max Eyth machine-tool factory in Stuttgart. As a student, Karl Benz visited the factory to see how the engine worked. It was his first encounter with the new technology. From then on, the internal combustion engine was to dominate his life.

The small, primitive Lenoir engine was a symbol of the great scientific and social changes taking place in Europe. More people were moving to the towns, to work in the new factories. New industries were growing and demanding power. A network of steam railroads was being built. Rail travel had proved that human beings could safely journey faster than a horse could gallop. Steamships were crossing the oceans, and steam propulsion had been tried in airships, though without great success.

Below *Steam technology was widely in use when Karl Benz was a student. This is a steamship, used for delivering mail.*

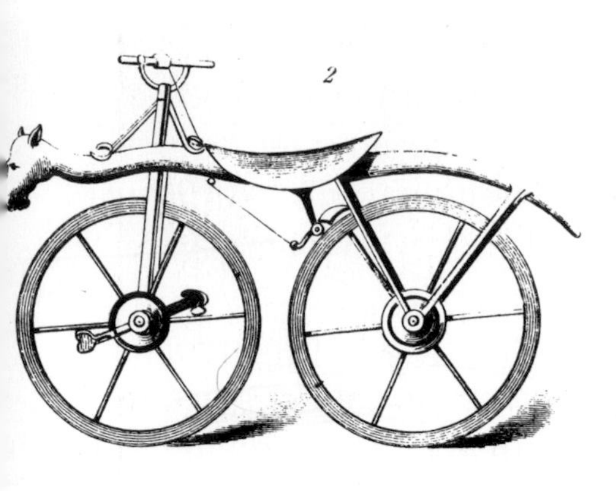

The "hobby-horse" was improved by adding pedals to drive the front wheel. The solid wheels and hard saddle made it an uncomfortable machine to ride.

For most people, transportation meant a horse, or their own feet. The first mechanized personal transportation was the two-wheeled hobby-horse, a machine first demonstrated in 1813 by its German inventor, Baron Drais. Hobby-horse riders sat astride the crossbar and pushed the machine along with their feet. In 1839, five years before Karl Benz was born, a Scot named Kirkpatrick Macmillan improved the hobby- horse, making it more like the bicycle we know today. Macmillan never developed his invention. The bicycle was re-invented in 1861 by the French engineer Pierre Michaux. The new machine, named the velocipede, became popular, especially among young people. As a teenager, Karl Benz shared the craze for cycling that swept Europe.

Karl Benz watched cyclists pedaling their way slowly along the rough roads. It was hard, slow work, especially uphill. In the workshop, he listened to the steady beat of the single-cylinder gas engine, and thought about the bicycle. Bicycle wheels could be turned by chains and cranks, using the power of human muscles. Could wheels also be turned, and with far greater speed and power, by the explosions of the gas-burning engine?

A modern BMX bike. Since the 1860s, the bicycle has evolved as a cheap and efficient means of personal transportation.

Steam on the Road

The first machine known to have moved under its own power on land was a large gun carriage. It was a three-wheeler, built in 1769 by Nicolas Joseph Cugnot (see illustration below). It was powered by a steam engine that drove the front wheel. Its working life was brief. After several slow test runs, the machine went out of control and overturned. It was locked away, before it could do any damage.

Other inventors, such as the American Oliver Evans and the Briton Richard Trevithick, experimented with steam vehicles. Trevithick built his first steam coach in 1801, naming it the "Puffing Devil." With some friends as passengers, he drove it a short distance and stopped for a meal at a roadside inn. While the travelers were busy eating, the boiler of the steam coach dried out, the engine overheated, and the "Puffing Devil" set fire to some outbuildings. In a second steam coach, Trevithick drove from Cornwall to London in 1803. The engine had seized up by the end of the trip. This was hardly surprising, considering the jolts it suffered over the appalling roads of the day. Trevithick concluded that the future of steam transportation was on rails, not on roads.

2 The Engineer and the Engine

As an engineering student, Karl Benz learned about steam engines. The steam engine was the driving force of the new machine age. Its power came from heating water to make steam, which then pushed a piston up and down inside a cylinder. The steam engineers had to solve the problem of converting the up-and-down, or reciprocal, motion of the piston into a useful turning, or rotary, motion. In the earliest factory steam engines, the piston was connected to a rocking beam. The beam was linked through a system of gears to a large drive wheel. Pulleys and belts ran from the drive wheel to the factory machinery.

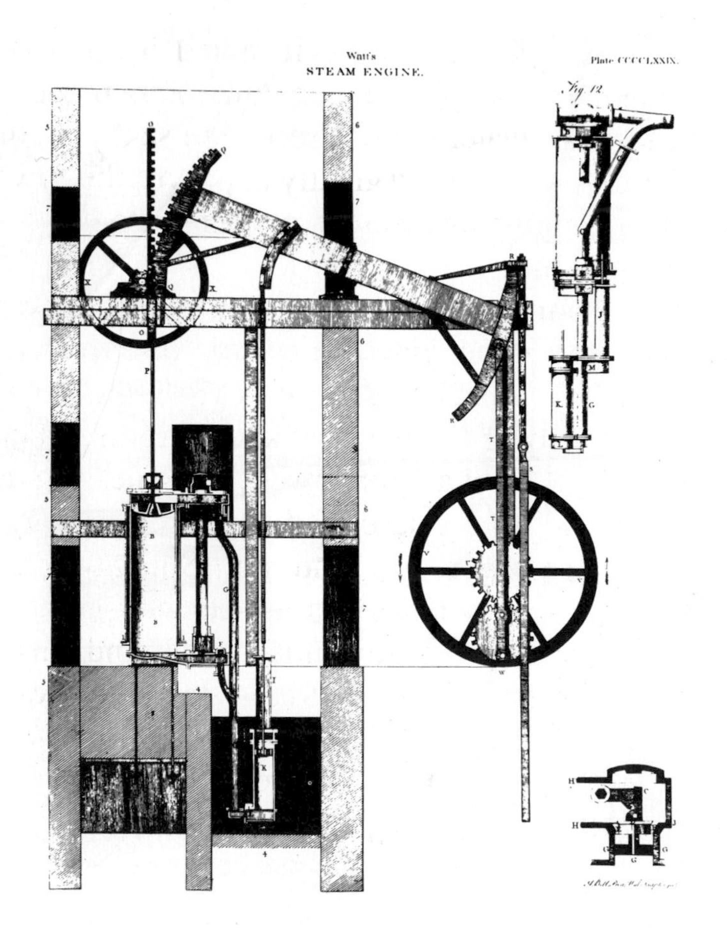

James Watt's improved steam engine. The first steam engines were built to pump water from mine shafts. Later, they were adapted for work in factories.

This steam omnibus provided a passenger service in London for a short period in the 1830s.

The steam engine had been invented in the early 1700s, when Thomas Newcomen built steam pumping engines to draw water out of deep mine shafts. In the late 1700s, James Watt had greatly improved the power of the steam engine by adding a condenser to increase the steam pressure.

Several components of the steam engine were adaptable into a gas engine: the piston, the valves and the toothed gear-wheels. As early as 1808, the British inventor Sir George Cayley, best known for his glider experiments, devised a "gunpowder" engine. Around 1820, scientific papers described engines driven by a mixture of hydrogen gas and air. Small engines powered by hot air were also built, but these produced only enough power to drive dentists' drills and fans.

Several attempts were made to use steam engines to drive road vehicles. The steam carriages worked well enough, but they faced three great problems: their greedy appetite for coal and water; a tendency to explode; and – most damaging – fierce opposition from railroad owners and other road users.

Making steel by the Bessemer process. After 1856, Sir Henry Bessemer's method for converting iron to steel made it possible to manufacture steel in large quantities.

So the search for an alternative engine continued, although with no great urgency. In the 1820s, Samuel Brown of Britain built a "gas-vacuum" engine, which burned hydrogen. The gas was burned inside the cylinder to create a vacuum. Atmospheric pressure then forced the piston downward. Brown fitted the engine to a carriage and drove up Shooters Hill in Kent, England, but he never developed his invention. Lenoir also tried using his gas engine to power a carriage.

In 1862, he managed a journey of 5.6 miles (9 km) in three hours. This unpromising test was not repeated; but it aroused the interest of Karl Benz. Watching the stationary Lenoir gas engine in the workshop, he observed that it had a tendency to shake itself to pieces. To work more smoothly and efficiently, a gas engine would need at least two cylinders. Many other problems still needed to be solved before the gas engine could hope to challenge the supremacy of steam.

In 1864, Benz graduated from the technical college, where his teachers had been impressed by his determination and hard work. His first job was not as an engineer but as a locksmith. This allowed him to stay near his mother, who was now a sick woman. His first engineering job was as a fitter in a large factory in Karlsruhe, which produced heavy machinery, including railroad locomotives.

In the factory, Benz worked a thirteen-hour day. It was hard, dirty work, and often dangerous. One day, Karl was working on a massive engine forging that was supported by wooden props. The props gave way, and the forging crashed to the ground, pinning Karl and another workman beneath the splintered wood. Fellow workers rushed to drag them clear, and fortunately neither was seriously hurt.

Karl Benz worked in a factory like this one, helping to make heavy machinery, boilers and railroad locomotives.

Several years as a factory fitter taught Karl much about practical engineering. In 1869, he moved to a new job, shortly before a new production director joined the Karlsruhe Maschinenbaugesellschaft, the factory where he had been working. The newcomer was Gottlieb Daimler. The two men, Benz and Daimler, never met. Yet their names were destined to be linked with the birth of the automobile.

In 1870, Prussia and France went to war. The result was defeat for France and the creation of a united German Empire under Prussian rule. In the spring of that war year, Karl's mother died, and at the age of twenty-five he faced the future alone.

For a time, he worked with a firm of bridge builders in Pforzheim. He met a girl named Berta Ringer and they married in July 1872. Karl was ambitious and Berta had some money of her own, so he set up his own business, with a small workshop and a partner. He soon bought out the partner, preferring to make all the decisions himself. But trade was poor, and little money came in. The birth of two sons – Eugen in 1873 and Richard in 1874 – added to the family's worries. Their financial position went from bad to worse. The bank rescued the firm with a loan that saved the workshop, but Benz had to sell all his tools to pay off his debts.

Otto von Bismarck (right) with the defeated French Emperor in 1870. Under Bismarck's leadership Germany was united under Prussian domination and trade and industry were encouraged.

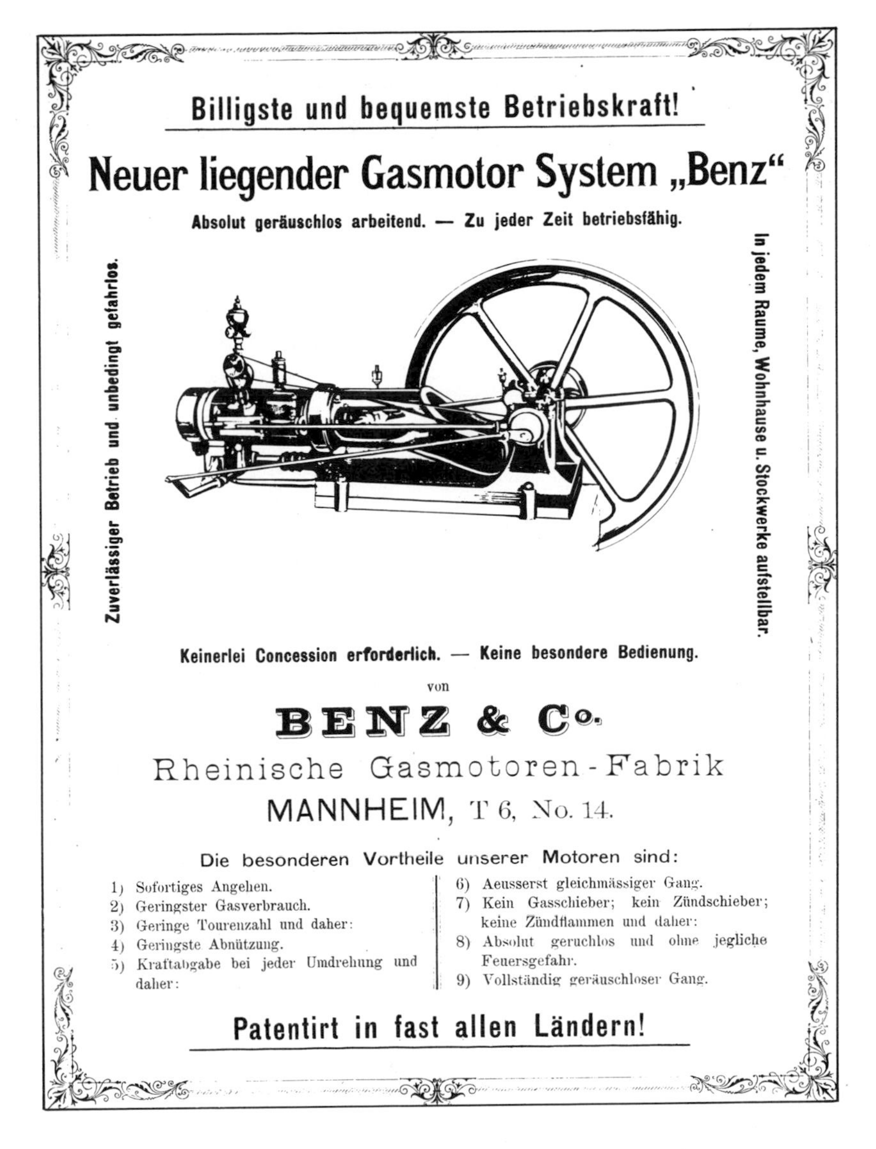

An advertisement for Benz's stationary gas engine. Benz hoped that this small but efficient engine would improve his fortunes.

From 1878 to 1880, Benz was at work designing and building his first gas engine. It was to be a two-stroke. Benz knew that Nikolaus Otto had already made a four-stroke engine, the design of which was protected by a patent. Benz could not copy it, so he had no choice but to develop his two-stroke engine. He spent days, and often nights, at his bench, because every part of the engine had to be designed and built by hand. The business suffered, and on New Year's Eve 1880, he and Berta had no money to celebrate. But they were happy; a steady "put-put" sound was coming from the workshop. The little two-stroke engine was finished and running.

3 Problems and Solutions

One of Dr. Nikolaus Otto's horizontal gas engines. Otto-type engines used coal gas from the municipal supply. They were suitable for factories, but useless for driving vehicles.

Benz had his engine, but he still had little money. He was not a good businessman, yet somehow he had to sell engines to build up the company. However, he knew that the two-stroke was not really the engine he wanted to build.

In 1862, a Frenchman, Alphonse Beau de Rochas, had designed a four-stroke engine in which the gas and air mixture would be compressed before ignition. Compression increases the engine's power, because the fuel burns more efficiently. The problem is the resulting heat – about 4,530°F (2,500 °C) inside the cylinder of a car engine. Without efficient cooling, the engine would melt. Beau de Rochas never built his engine.

The two-stroke is simpler than the four-stroke engine (the stroke is the up-and-down movement of the piston). The four-stroke has intake, compression, power and exhaust strokes. The two-stroke combines exhaust and intake near the end of the power stroke. A modern gasoline lawnmower has a two-stroke engine. It is cheap, and gives more power than a four-stroke of the same size. This is because in a two-stroke, the engine produces a power stroke for every turn of the crankshaft. In the four-stroke, there is a power stroke for every other turn of the crankshaft.

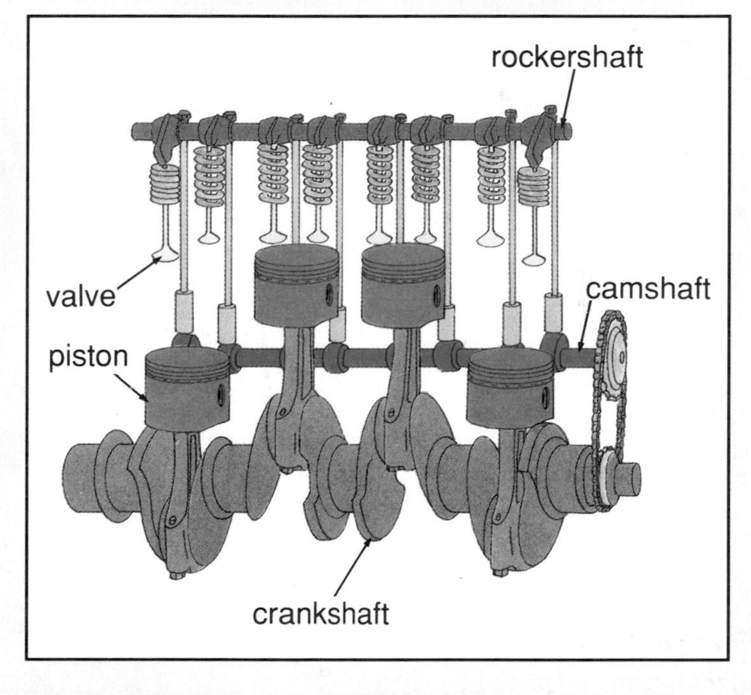

A diagram of part of a four-stroke engine. The crankshaft is turned by the movement of the piston. The crankshaft's movement is transferred to the camshaft by the drive chain, on the right-hand side of the diagram.

Nikolaus Otto's "silent gas engine," patented in 1877, was the first working four-stroke engine. Otto's partners were Karl Langen and Gottlieb Daimler. Before joining the Otto and Langen company, near Cologne, in 1892, Daimler had traveled and spent some time in England. Otto was happy to sell the new engine as a stationary power unit. Daimler believed that it could power a vehicle, but that it would need a different fuel and an ignition system that did not rely on a permanent gas jet.

Unknown to Otto and Daimler, Benz was already working on both problems. In the 1850s, oil had been discovered in Pennsylvania. The growing United States oil industry supplied fuel for use in oil lamps, as well as the heavier types of oil that were used for lubrication. However, no one had yet found a use for the lighter "fraction" of the oil-refining process – petroleum spirit, or gasoline.

In 1870, an Austrian named Julian Hock made a two-stroke engine that worked on gasoline. Benz decided to burn gasoline in his engine. He set to work designing a carburetor, in which the fuel and air would be mixed to produce an explosion inside the cylinder. To ignite the explosion, there must be heat. A spark was the neatest solution, but it was a challenging puzzle to devise an electrical ignition system that would work reliably. There were many other problems, such as cooling and lubrication, to keep Benz at his workbench night after night.

An oilfield in Pennsylvania, in 1871. The new oil industry provided an unexpectedly useful by-product – gasoline.

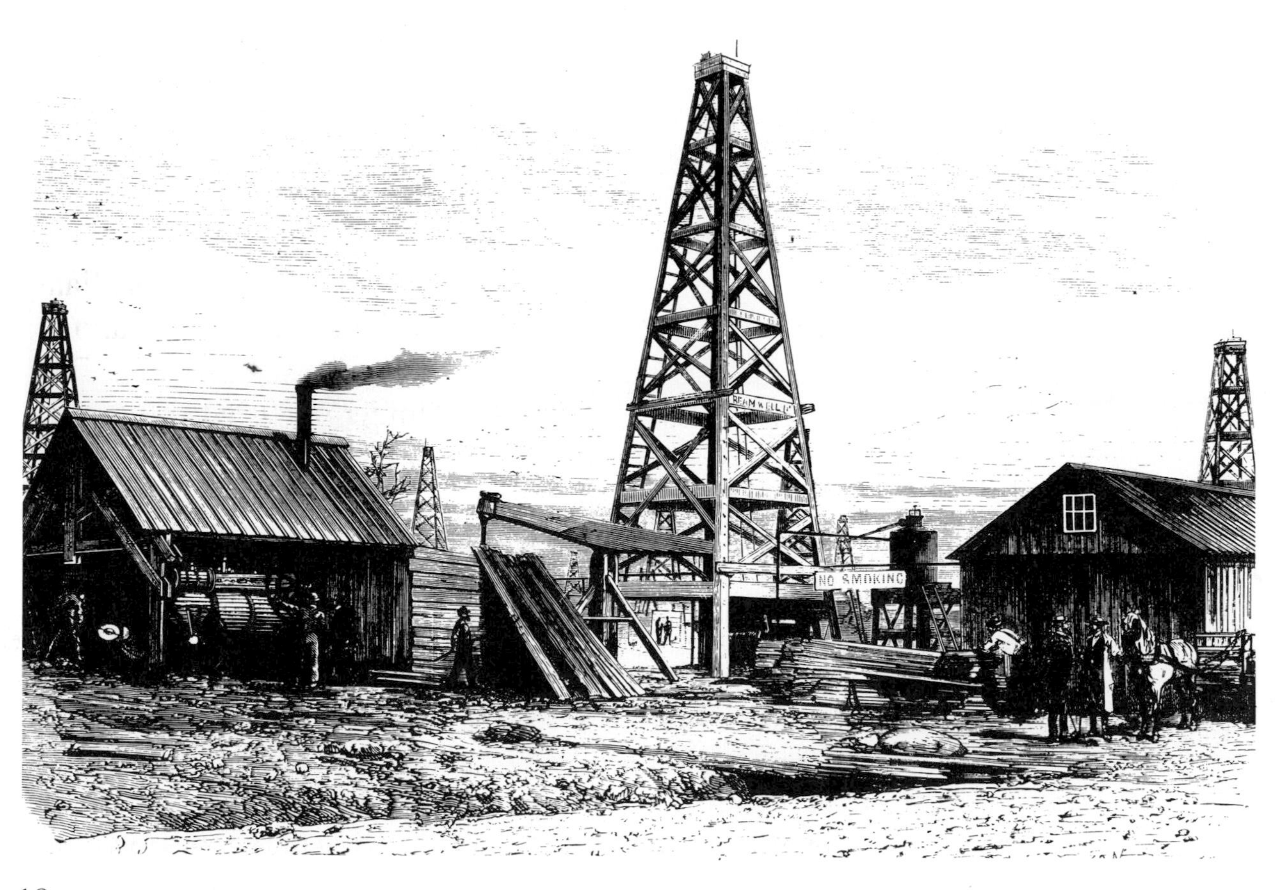

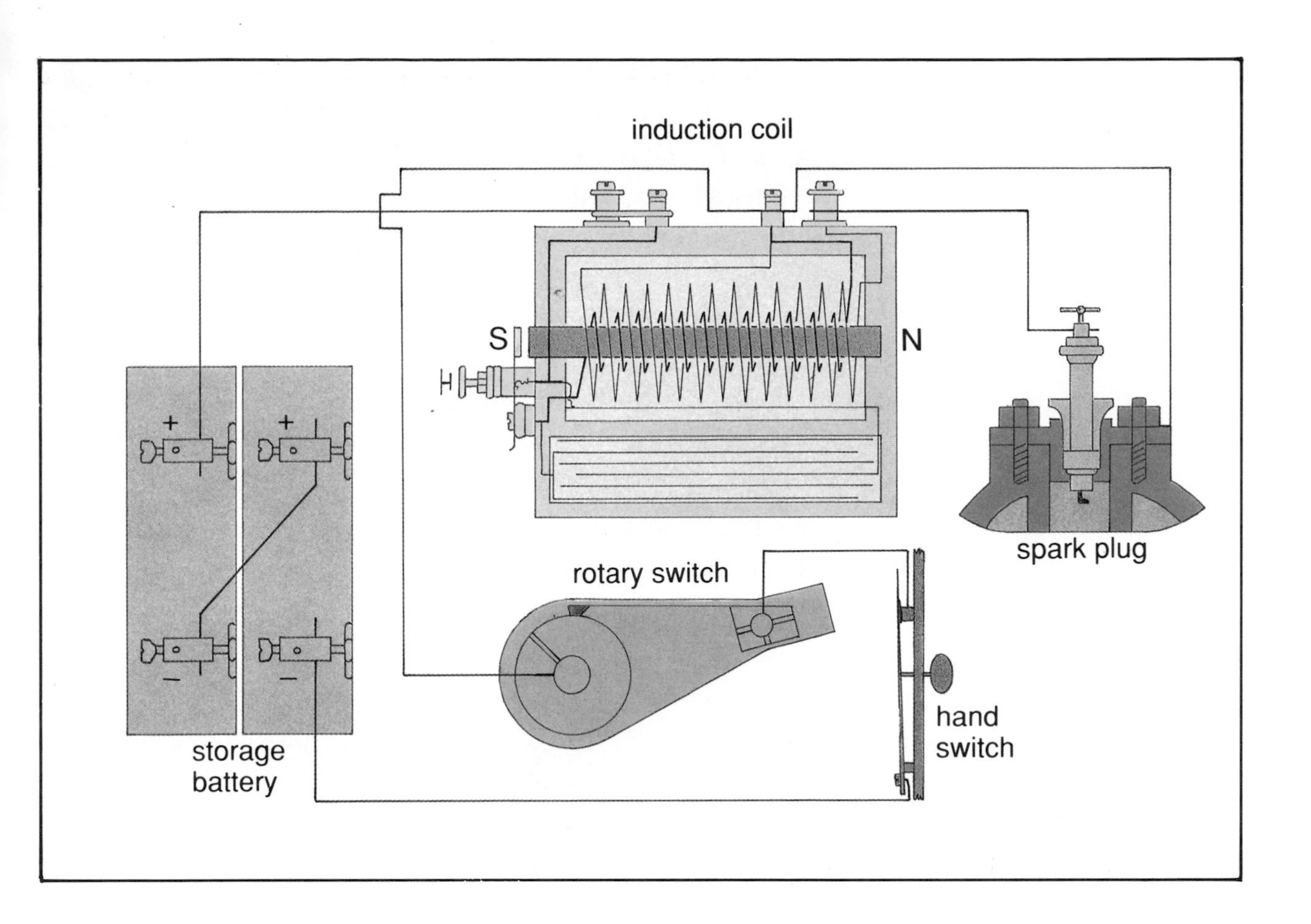

A diagram showing Benz's electrical ignition system. The battery supplied power to a coil, which sent a current sparking across a gap in the spark plug. The spark ignited the mixture of gas and air inside the cylinder.

It is never easy for a pioneer to remain enthusiastic about a new invention when public interest is very low. Julian Hock gave up and is forgotten. So did another Austrian, Siegfried Marcus. In 1873, he built a single-cylinder gasoline engine. Marcus was a brilliant innovator. He designed an ignition system and a carburetor in which a revolving brush sprayed a fine mist of gasoline into the cylinder. He tried the engine in a wooden carriage. Tests carried out on a rebuilt model in 1949 showed that it produced ¾ hp (horsepower), giving a top speed of 5 miles (8 km) per hour. But Marcus lacked Benz's determination and turned to other things. He missed his chance to go down in history as the inventor of the automobile.

Benz now had some good fortune – finding a new financial backer, a photographer named Emil Buhler. He began to sell some engines, and in 1881, his business was officially registered as the Mannheim Gas

Motor Works. Money was still short. The Benz family now included four children, and to add to their difficulties, Karl undertook and lost a costly lawsuit with his ex-partners. He left the Gas Motor Works and in 1883 he was on his own once more. Once again, help arrived. This time, two friends named Max Rose and Friedrich Esslinger put up the money to start a new Benz company.

Free from money worries, Benz devoted all his energies toward a new goal. He was determined to build a motor vehicle. The spur he needed came with the good news that the patent protecting the Otto four-stroke engine was no longer in force. He was free to build his own four-stroke, which would drive his revolutionary "gas-engined carriage."

Opposite *Nikolaus Otto pioneered the four-stroke engine cycle, which is used in almost all modern car engines. A stroke is one movement of the piston from one end of the cylinder to the other.*

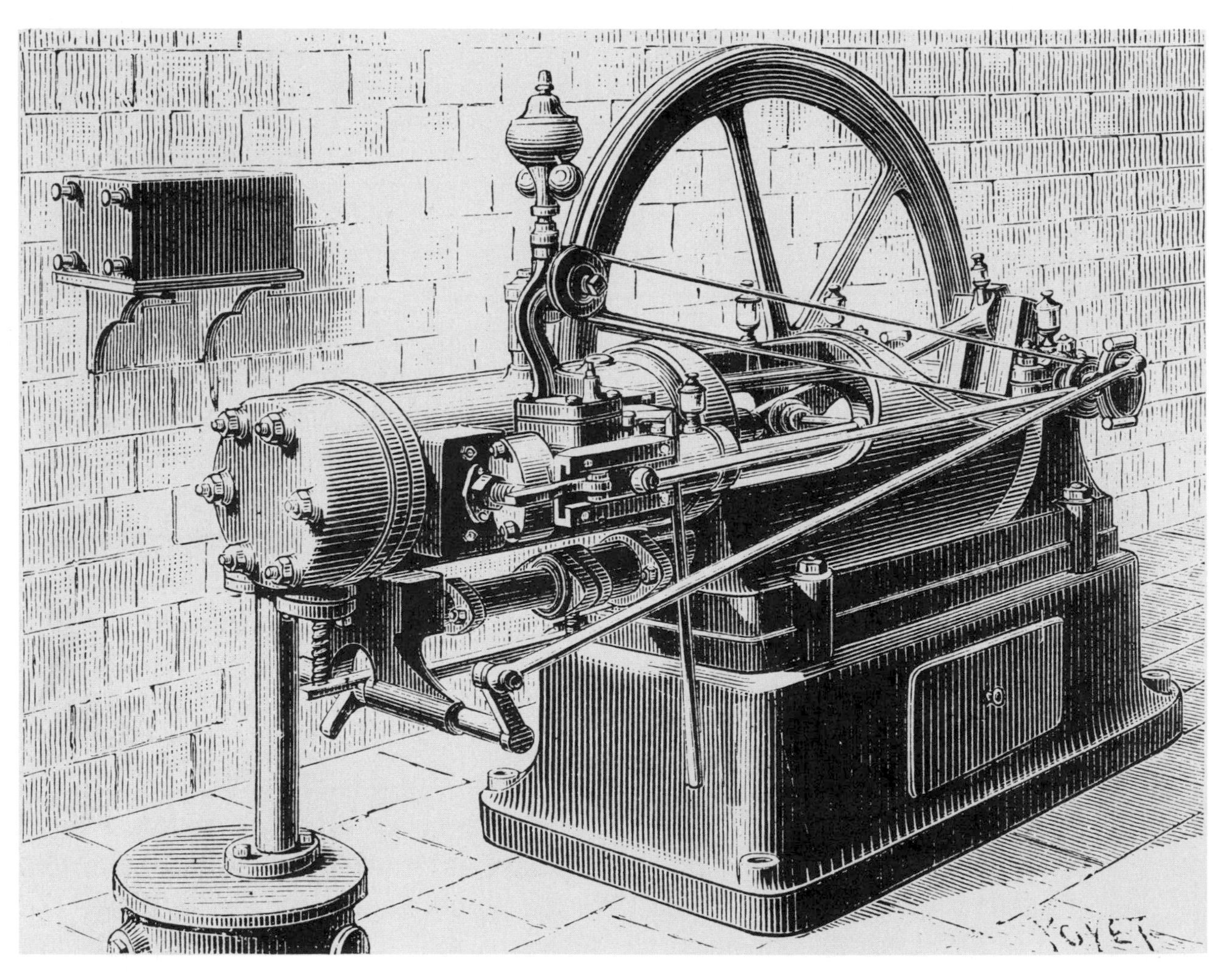

Below *A Benz gas engine of 1886. Working with stationary engines, Benz developed the ignition and other systems he built into his first car engine.*

The four-stroke cycle

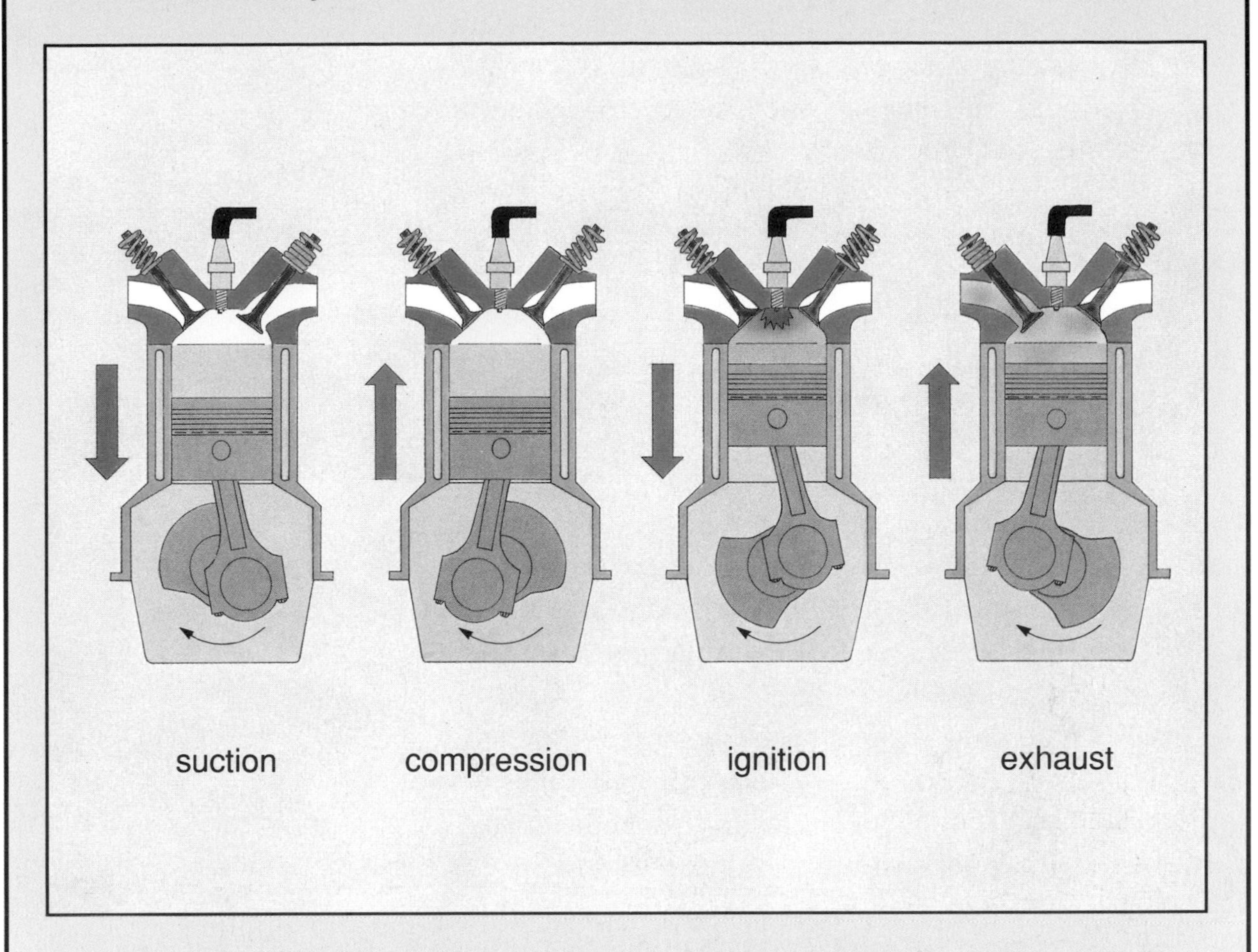

Nikolaus Otto's engine was an improvement on that of Lenoir. The explosion of the gas drove a free, heavy piston upward. As the gas cooled, atmospheric pressure plus the piston's own weight forced the piston downward to engage and rotate the drive shaft.

The engine was noisy and heavy, and Otto improved it in 1876. It worked on the same four-stroke ("suck, squeeze, bang, blow") cycle we know today. This is how it works.

1. On the **suction** stroke, the gas/air mixture is sucked into the cylinder through an inlet valve as the piston moves downward.
2. On the **compression** stroke, the piston travels upward, squeezing the mixture into the space above it.
3. On the **ignition** stroke, the power of the engine is generated by the explosion of the gas/air mixture, ignited by a hot spark. The gas expands as it is heated and the piston is forced downward.
4. On the **exhaust** stroke, the piston moves upward and pushes the spent gas out through an exhaust valve.

4 On The Road

Benz set out from the start to build a new kind of vehicle. Other inventors had installed gasoline engines in wooden carriages that could just as well have been fitted with shafts and pulled by horses. Benz built an automobile.

He had to work out a shape for his vehicle and the systems to make it work. He had to design and build a carburetor and to choose an accumulator (battery) that would produce enough electrical power to make a spark at the plug. He also had to design a method of cooling the engine. To do this, he enclosed it in a water-filled jacket, linked to a simple radiator that would release heat to the air.

Benz decided that the car must be three-wheeled. This would not need such a complicated steering arrangement as a four-wheeler. Benz copied the design of the pedal tricycle, providing his car with a single unsprung front wheel steered by a lever. The rear axle, with its two large wire-spoked wheels, was mounted

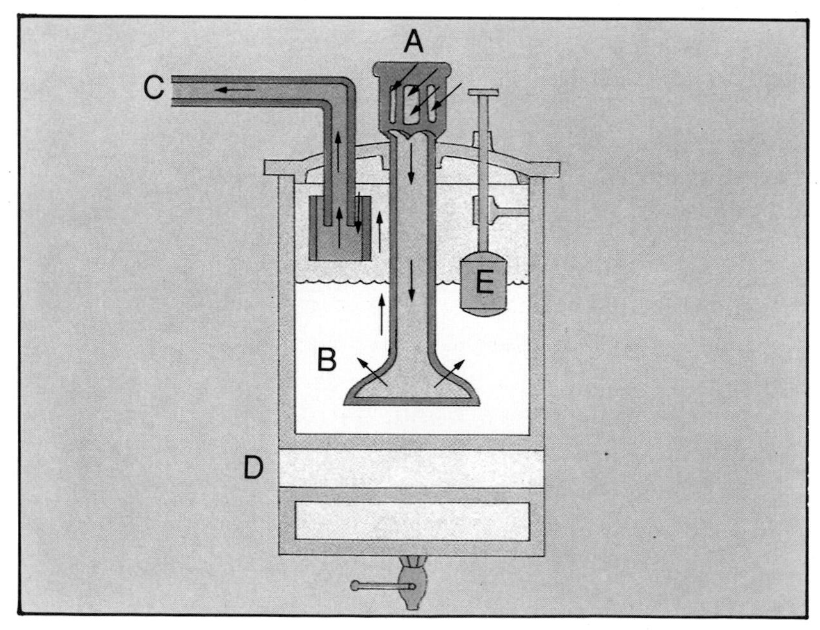

In Benz's carburetor, air was sucked in through inlet A, and passed through the gasoline in B. From there, it moved on to the engine at C. The gasoline was warmed at D, using heat from the exhaust gases. The gasoline level was controlled by the float at E.

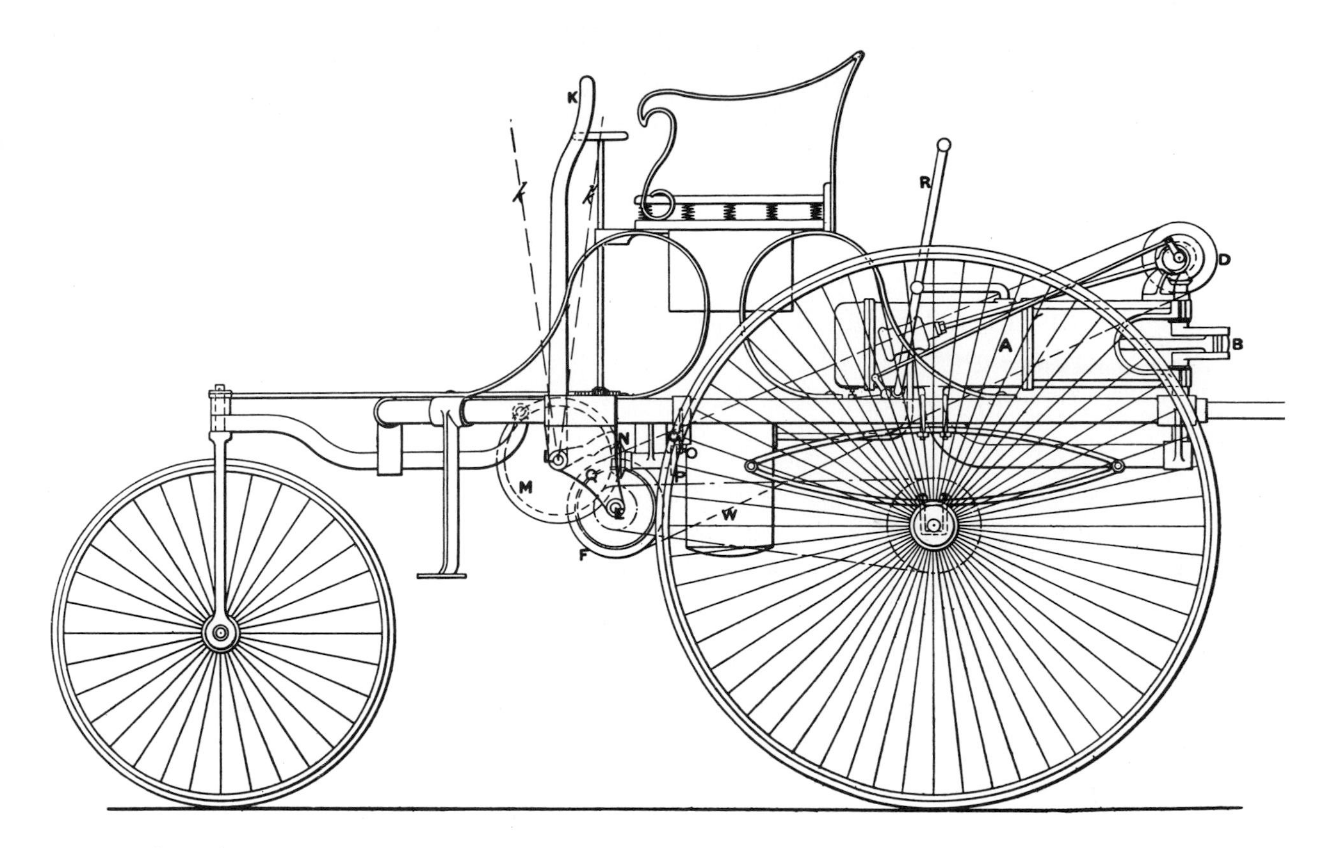

on metal springs.

The driver and passenger sat on a bench seat. Behind and beneath them was the heart of the car: the engine. Benz's first engine had a capacity, or internal volume, of about 0.42 gallons (1.6 l). This is the same size as the engine of an average family car today. However, it was far less powerful.

The Benz engine turned very slowly, at around 250 revolutions (turns) per minute (rpm). The speed of a modern car engine is measured in thousands of "revs." The engine weighed 209 pounds (95 kg) and was mounted horizontally, so the piston was on its side (unlike those in a modern car engine). A large flywheel was also set horizontally. The flywheel's motion stored power from the engine, so that the engine continued running throughout the complete four-stroke cycle. The engine was started by spinning the flywheel.

The trickiest problem for Benz was the transmission – linking the engine to the drive wheels through a gear arrangement. He needed a differential gear, so that the

Side view drawing of the 1885 Benz tricycle. The long gear lever is to the left of the driver. The design of the first car was closely related to the bicycle.

wheels would turn at different speeds when the car was traveling around a bend. The differential gear he used was a recent invention. It had been patented in 1877 by the British engineer James Starley, originally for use on pedal tricycles.

Front view of the tricycle. The rear-wheel drive chains can be seen, and also the driver's tiller that turned the single front wheel. The tricycle's frame was made from metal tubing. This historic vehicle had a top speed of about 10 mph (16 kph).

Back view of the 1885 tricycle, showing the engine with its large horizontal flywheel. The water tank was mounted above the single cylinder. The fuel tank was on the left side of the engine, beside the cylinder. The rubber-tired wheels had wire spokes.

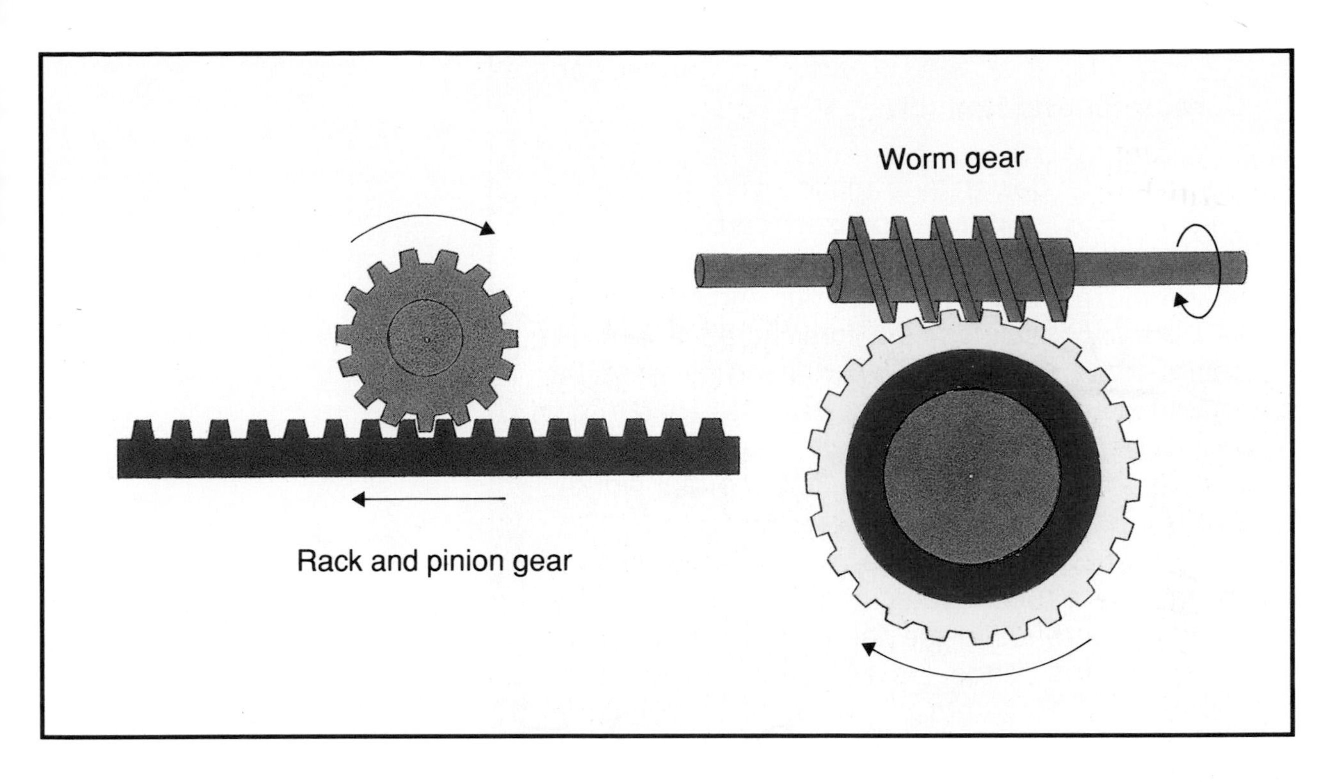

A diagram of two common types of gears. Gears had always intrigued Karl Benz. Inside a modern gearbox are arrangements of toothed wheels. Benz's first car had a gear system using pulleys and belts. To put the car into gear, the driver pulled the lever backward. Pushing it forward applied the brake.

Gears had fascinated Benz ever since, as a child, he had amused himself by taking clockwork mechanisms to pieces. He recalled how much he admired "the marvelous language that gear wheels talk when they mesh with one another."

Benz based the transmission on the belt-and-pulley layouts he had seen many times in factories. There was no pedal-operated clutch (as in modern cars) to disconnect the engine from the drive wheels. Instead, Benz used an ingenious system of moveable (or loose) and fixed pulleys. When the transmission belt was shifted from the loose pulley to the fixed one, the clutch effect was obtained and the engine was disconnected from the wheels.

For the final drive, Benz turned again to the bicycle for his inspiration. He linked the shaft to the drive wheels by side chains. The car was fitted with brakes; these were rather crude blocks that were applied not to the wheels but to one of the pulley wheels in the transmission.

Though small and crude by modern standards, Benz's first car still represents an astonishing technical

Carburetor and ignition

Benz had to design and make every part of his engine. His surface-vaporizer carburetor was simple but bulky. Air was sucked in through an inlet valve and passed over a porous (holed) surface kept wet by gasoline in the float chamber. Exhaust gases from the engine were used to heat the chamber, so that the gasoline vaporized and mixed with air before it entered the cylinder. The result was a weak and variable mixture, but one that worked well enough in such a small engine.

Daimler solved the "spark plug" problem by using a metal tube, heated from the outside, so that it glowed red-hot inside the cylinder head. Benz designed and made his own spark plug. In his ignition system, electricity from an accumulator, or wet battery, was passed to an induction coil, which in turn produced a spark between the gaps of the plug.

breakthrough. Four features of the car became standard on later versions: the water-cooled engine, the electrical ignition, the differential gear and the mechanically controlled valves that were opened and shut by the camshaft.

In the spring of 1885, Benz had the car ready for testing. On the first trial, Berta watched excitedly as Karl drove the machine slowly around the factory yard. On the second run, she climbed onto the seat beside him. The flywheel was spun, but the engine refused to fire. Benz tried again, and this time the engine spluttered into life and the car leaped forward, crashing into a brick wall.

The front forks were broken, but quickly repaired. Neither driver nor passenger was hurt, and they were soon back on the road, venturing about 110 yards (100 m) before the engine died and the car came to a halt. Each run taught Benz more about the workings

Benz's little three-wheeler was a breakthrough in transportation technology: the first practical automobile to be powered by a gasoline engine.

of the new machine. Each time they traveled a bit farther, until the car was running for over half a mile.

In the summer of 1886, Benz took the car out onto the roads. It managed distances of more than half a mile, at a speed of about 7.5 mph (12 kph).

Journalists now took an interest in the new machine. On July 3, 1886, the local newspaper described how "a velocipede driven by gas was tested early this morning on the Ringstrasse, during which it operated satisfactorily."

The car created a sensation in Mannheim. Crowds gathered to watch it approach. Horses shied in terror. One old lady fled indoors, convinced that Benz was the devil, driving an infernal carriage. There was such a commotion that Benz received an official warning from the local police to curb his test drives. To escape public attention, Benz took to testing the car at night.

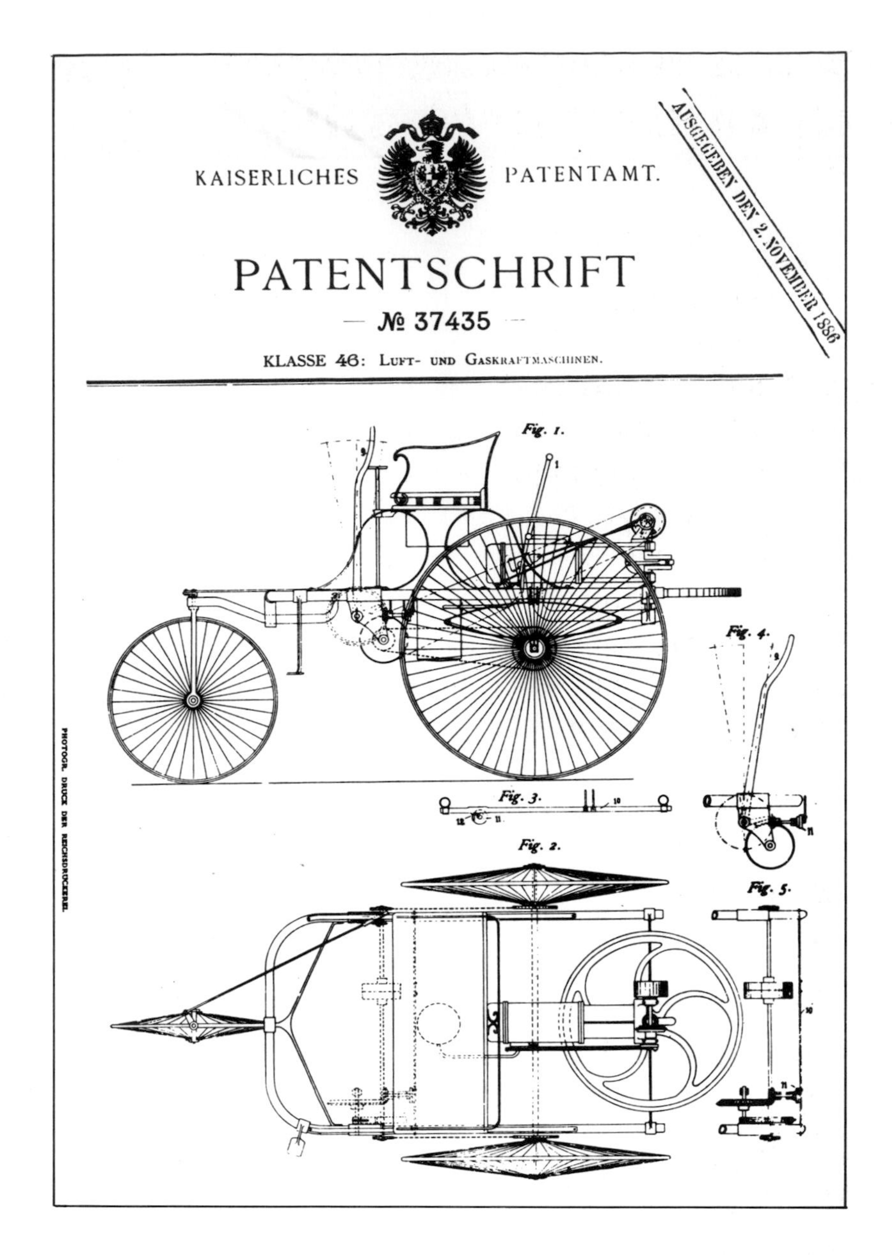

KAISERLICHES PATENTAMT.

AUSGEGEBEN DEN 2. NOVEMBER 1886.

PATENTSCHRIFT

— № 37435 —

KLASSE 46: LUFT- UND GASKRAFTMASCHINEN.

Fig. 1.

Fig. 2.

Fig. 3.

Fig. 4.

Fig. 5.

PHOTOGR. DRUCK DER REICHSDRUCKEREI.

Benz's first patent, No. 37435, was awarded to him in November 1886.

Benz was delighted by the success of his little three-wheeler, but his business partners were critical. Karl's obsession with the automobile kept him from his work on the stationary gas engine. What was the use of his invention, they argued, if no one was willing to buy such a machine? Why should anyone buy a vehicle that was no quicker than a horse and broke down or quickly ran out of fuel?

Benz had produced a great technical success, but he still had to convince the world that the automobile was here to stay.

5 Winning over the Public

Karl Benz was not a man to give up once he had set his mind on a goal. He was convinced that his "horseless carriage" had a future and began manufacturing cars for sale – the first person in the world to do so. He received total support from Berta. The Benz household must have eaten, drunk and dreamed automobiles in those early days. At night, Berta sat at her sewing machine, but she was not dressmaking. The machine's foot-operated treadle pulley was linked to a small dynamo, which recharged the accumulator that provided the electrical power for the car's ignition.

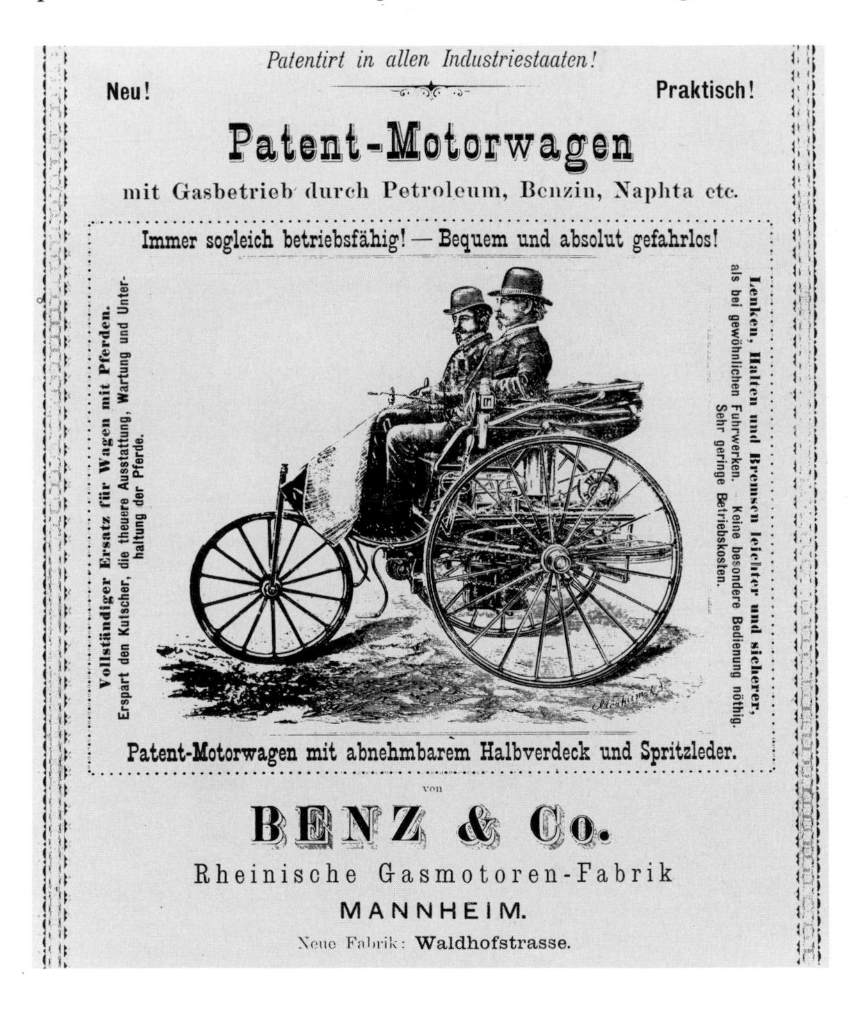

Benz advertising in 1888 stressed the comfort and practicality of cars.

In 1887, Benz traveled to the Paris Exhibition, but was disappointed by the lack of interest from the public. He had improved his car considerably. The engine was more powerful, and the front of the body was mounted on springs, although the front wheel was still unsprung. The brakes were better and the bodywork more comfortable. An improved two-speed gear made starting and hill climbing easier.

Benz was very annoyed to read in the Mannheim newspaper that his invention was "useless, ridiculous, and indecent." Why, asked the newspaper, should anyone wish to spend money, and risk life and limb, in the purchase and use of an automobile, when there were still plenty of horses for sale?

Fortunately, not everyone shared this hostile opinion. The automobile was beginning to attract supporters. One convert to the idea of "motoring" suggested that it would be ideal for country doctors. A doctor could drive to an emergency call without the delay caused by rousing his servant to harness the horse and carriage.

Among the first motorists were doctors, who needed fast transportation to reach their patients. Modern ambulances are vital for today's medical services.

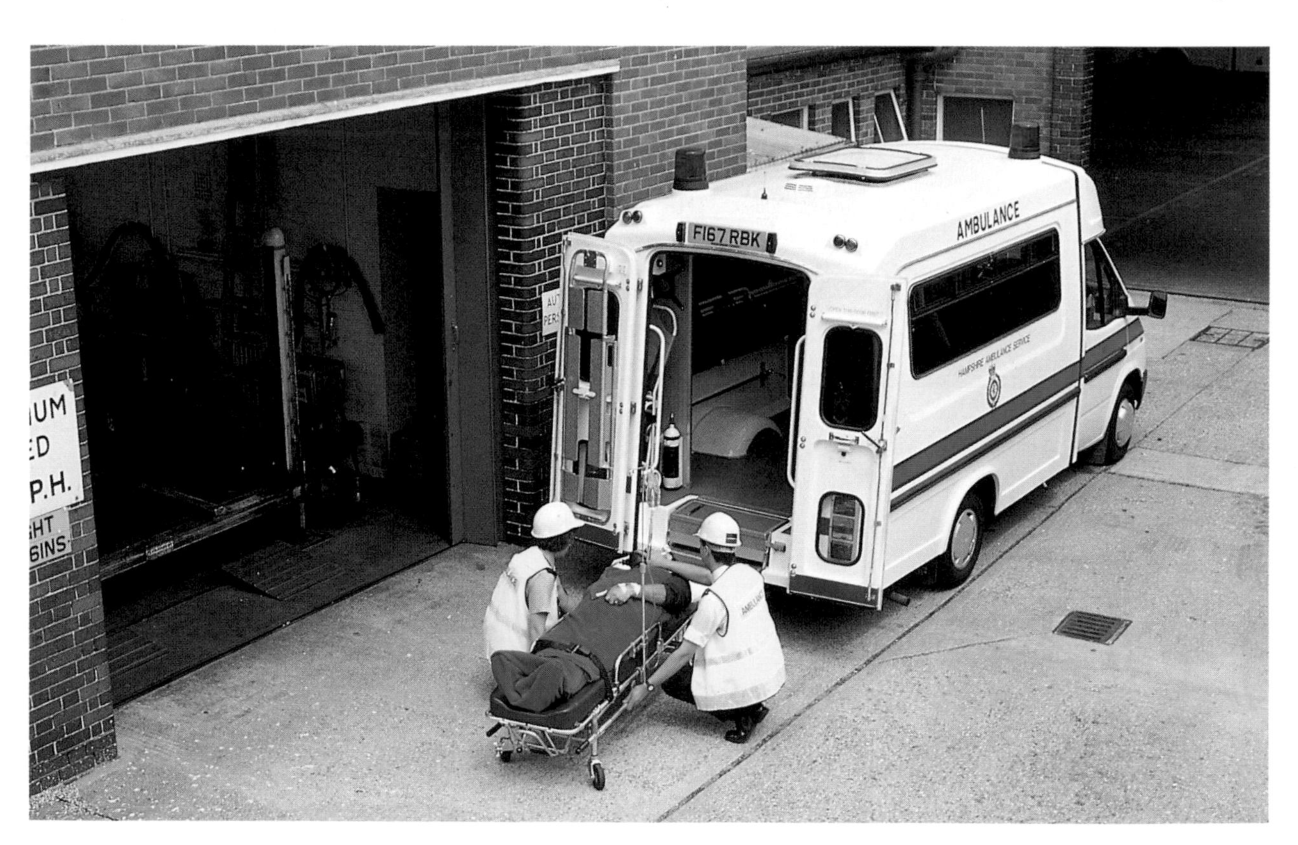

Berta Benz's drive to Pforzheim in August 1888 was the first long cross-country trip in motoring history. With her sons, she stopped to buy gasoline from a druggist.

Berta Benz was equally determined to show the importance of Benz's invention. One morning in the summer of 1888, she got up very early while Benz was still in bed, and took their sons Eugen and Richard for a drive to Pforzheim. The roads were dusty, stony tracks, intended for horses, not cars. There were, of course, no filling stations. Berta bought gasoline from the local druggist. Going uphill, she and Eugen got out and pushed while Richard steered.

Other road users must have been startled to come face to face with the first car they had ever seen, as it made its slow journey across country. At one village, Berta sought help from the cobbler, who hammered fresh leather onto the wooden brake block. She used one of her garters as insulating tape to mend a short circuit in the electrical system, and cleared a blocked fuel jet with a hairpin.

The first appearance of Benz's car on the roads aroused great interest and excitement.

The journey to Pforzheim took all day. Benz, who must have been extremely proud of his wife's achievement, sent her a message. He asked her to send back the drive chains on the first fast train, since he needed them for a new car being prepared for the trade fair in Munich.

At Munich, Benz persuaded the exhibition organizers to let him drive the car through the streets. They agreed, on condition that he pay for any damage to people or property. The car's public demonstration made an exciting news story, and the newspapers gave the Benz car plenty of publicity. Karl was awarded a special gold medal.

Despite this acclaim, the Benz cars were not selling very well. Few people were willing to put their money, or their lives, at risk. Also, there was a never-ending stream of official forms and regulations, such as the one that insisted that a bell be rung loudly to warn pedestrians of the car's approach.

Benz spent most of his time in the workshop. Almost every part of the car was made under his personal supervision. In 1889, he went to the Paris World's Fair, hopeful of success. However, the only way to sell Benz cars in France was to set up a local agency with a French partner. By now, Benz's German partners had run out of patience and money. They refused to give any more financial support and left the company.

Once again, at a critical time, Benz was fortunate in finding friends. Friedrich von Fischer and Julius Ganss suggested a new "rescue package." Von Fischer would look after administration, Ganss would be sales manager, and Benz would be free to concentrate on the engineering.

It was an ideal arrangement, for Benz's new partners had the skills he lacked. At long last the sales began to come in; not just from German customers, but from France, Britain, Belgium, Russia, Austria and even South America. Even so, the main business of the Benz company continued to come from the sale of stationary gasoline engines. These were in regular demand from factories in small towns with no municipal gas supply.

By now, Benz was not alone in making automobiles: Gottlieb Daimler was also in the contest. For some years, he had been working along similar lines to Benz. Daimler was an equally gifted engineer. But he was more widely traveled than Benz and knew more people in the business world. He was not easy to work with, but he had a brilliant assistant: Wilhelm Maybach.

Gottlieb Daimler was chiefly interested in engine design and improvement.

In 1872, Daimler had been appointed technical director at the gas engine works of Otto and Langen. For a time, Daimler and Otto worked together. However, Daimler was impatient and often tactless, and kept having disagreements with Otto. Eventually, after one particular argument, Otto fired Daimler.

With Maybach as his assistant, Daimler set up his own engine business in 1882. Daimler began work on an engine suitable for driving a road vehicle or a boat. His workshop was the summer-house in his garden at Cannstatt. He built three gasoline engines, ignited by a "hot-tube" ignition system of his own design.

The world's first motorcycle (right), built by Daimler in 1885. It had a top speed of 7.5 mph (12 kph), compared with its modern descendant (below), which can reach speeds in excess of 150 mph (240 kph).

Unable to afford a carriage with which to experiment, Daimler decided instead to install his gasoline engine in a two-wheeled vehicle. He built the world's first motorcycle, a bulky wood and iron contraption with a ½ hp engine. His son Paul drove it for 2 miles (3 km) in November 1885.

In 1886, Daimler and Maybach fitted a larger one-cylinder engine of 1 hp to an ordinary four-wheeled horse-carriage. The engine drove the rear wheels. Daimler's car managed a top speed of 17 mph (28 kph). Which inventor – Benz or Daimler – deserves the credit for being the "father of the automobile"? Both took Otto's gas engine as their starting point. The essential difference is that while Benz set out to build an automobile, Daimler was more interested in a basic power unit, which could be attached to any kind of vehicle or could drive machinery. Daimler used his first car as a means of testing his engine. Unlike Benz, he did not investigate the various systems necessary to develop the automobile, such as gears, transmission,

The "red flag" laws

As the railroads spread across Europe and the other continents, few people were interested in an alternative road vehicle. In Britain, a few enterprising inventors, such as Sir Goldsworth Gurney and W. H. James, built steam coaches for road use and attempted to run passenger services. The engines were reliable, but the terrible roads took their toll. So did the steam coaches' enemies – the railroads and the horse-owning classes. Harsh penalties were imposed on horseless road vehicles. They were restricted to a maximum speed of 10 mph (16 kph) in the countryside, and to 5 mph (8 kph) in towns. In 1865, the "red flag" act was passed. By law, any self-propelled road vehicle had to be preceded by a man with a red flag (or a red lamp at night). The maximum speed was cut to 4 mph (6 kph) in the country, and 2 mph (3 kph) (slower than walking pace) in towns. By 1867, the steam coaches had been forced out of business. These laws effectively halted the development of road transportation in Britain for 30 years.

By the early 1900s, cars were becoming more powerful and more comfortable – like the Benz Phaeton limousine (above). The driver is still exposed to bad weather but the passengers sit in comparative luxury. These large cars had four-cylinder engines of about 40 hp, considerably less powerful than the six-cylinder engine of a modern Mercedes (right).

suspension and steering. Daimler's hot-tube ignition consisted of a platinum tube inserted into the cylinder and heated to red heat by a gasoline burner flame from the outside. Benz's spark plug electrical ignition system was much more sophisticated because it ensured the precise timing of each explosion within the cylinder. This is the system used today.

Daimler lost interest in automobiles for some years after his success of 1886. He designed gasoline engines for motorboats, airships and streetcars. He fitted his engines to taxicabs and fire engines, and offered them for sale as electricity generators.

Benz concentrated on automobiles. He was, without doubt, the first person to make cars that were reliable enough to be sold to the general public. From the 1890s, Benz concentrated on technical refinements. There was to be a constant battle between Benz and his business partners, who were interested only in making money from his invention.

6 Auto Progress

Benz turned his attention to the problem of powering four wheels rather than three. In 1893, he put his first four-wheeler on the road. It had an engine capable of producing over 3½ hp, and was called the "*Vis-a-Vis*" (Face-to-Face), because the passengers sat facing the driver.

In 1888, Benz abandoned the large horizontal flywheel and sct it upright instead. Later, he introduced starting handles on his cars. The luxurious "Victoria" and the cheaper "Velo" (1893) were his first really successful models. Both were single-cylinder machines, as was the eight-seat "Phaeton" of 1895. Benz was pressed by his partners to develop more powerful, faster engines, but he was reluctant.

The Benz "Velo Comfortable" of 1895. Although its top speed was only 17 mph (27 kph), it was reliable and sold well. The engine still had only one cylinder.

On the road

Benz named his "Victoria" car not after Britain's queen, but to celebrate victory over the difficult problem of designing an accurate steering system in a four-wheeled car. The "king-pin" (on which the front axle swivels) had been invented in the early 1800s by an artist-inventor named Rudolf Ackermann. It became standard on cars. The "Victoria" was the first Benz car made in quantity. It had a 3-hp engine and a top speed of 15.5 mph (25 kph).

Every drive for the pioneer motorist was an adventure. The authorities were alarmed at the prospect of speeding, and drivers were warned to slow down or even stop when passing or meeting a horse-drawn vehicle. Every hill was a challenge. Early cars had an extra brake that acted like an anchor, digging into the loose road gravel to keep the car from rolling downhill if the engine gave up the struggle!

In 1897, he produced his first twin-cylinder engine. By now, Benz cars were seen on the roads in Britain and France, and by 1898, Benz had sold a total of about 400 cars in those countries. Automobiles were particularly popular in France, which had Europe's best roads, and unlike Britain, had no restrictive road regulations. The French auto industry began with the

The eight-seater Benz Phaeton. Richard Benz is seated at the front nearest the camera. Karl Benz is on the back seat, between von Fischer and Ganss.

licensing in 1890 of the firm of Panhard and Levassor to build Daimler engines. René Panhard and Emile Levassor built their first car in 1890. Levassor's second car (1891) introduced a number of important improvements. The engine was moved from beneath the driver to the front of the car and was hidden under a square hood. The engine drove the front wheels. The car had a shaft-and-gear transmission and a friction clutch, allowing the driver to disconnect the engine from the drive shaft to the wheels.

In 1895, a Panhard-Levassor car came first of eighteen entrants in the world's first auto race, from Paris to Bordeaux. Benz disapproved of auto racing. He always drove very carefully and slowly, stopping the car if a horse-drawn vehicle approached. Racing on public roads, with its noise and danger, horrified him, and for years, he refused to allow Benz cars to race. His cars continued to be built along similar lines, even though manufacturers copying the "Panhard system" were beginning to make better models.

Above *One of the enduring creations of auto advertising: the Michelin man.*

Like Benz, Levassor fitted solid tires to his cars. Neither company used the pneumatic (air-filled) rubber tires first made by the Michelin brothers in 1895, since they punctured continually. But, as road surfaces improved, the air-filled Michelin tires offered much smoother running, so they rapidly became standard. Benz finally began using them in 1901.

By 1896, Panhard-Levassor was making cars with steering wheels (instead of levers) and four-cylinder engines. In 1898, Renault introduced the propeller shaft, replacing the bicycle-type chains that had been used on all the early automobiles.

By the 1890s, the United States had developed an interest in automobiles. In 1893, Charles and Frank Duryea built the first gasoline-driven car in the United States. In Britain, a few pioneers imported cars from Europe, and the first auto shows (in Tunbridge Wells and London) were held in 1895. When the campaign to do away with the old-fashioned "red flag" law finally triumphed in 1896, and the act was repealed by Parliament, an automobile parade was held from London to Brighton.

***Above** Crowds jammed Broadway to watch the start of the 1908 New York to Paris race.*

***Opposite** Automobile racing encouraged popular enthusiasm for the car. Challenge races were staged between rival models, as in this 1903 duel in Detroit, between Ford and Harkness cars.*

By the turn of the century, it was clear to most people that the automobile was here to stay. The managers of the Benz company saw that a great deal of money could be made and were eager to move into mass production. Already, the United States had taken the lead in the factory building of automobiles. Most European cars were still handmade, one at a time, like carriages.

In 1901, Ransom Eli Olds introduced the little U.S. Oldsmobile of 1901, which was the first factory-made production car. Five years earlier, in Detroit (later to become the center of the U.S. automobile industry), Henry Ford had made his first two-cylinder, 4-hp car. Ford believed that European car-makers were interested only in wealthy customers, since they made large, elegant and powerful cars that were very expensive. An example was the first Mercedes, built in 1901 by Maybach.

Ford decided to build a car for the masses. It had to be cheap, rugged and reliable. The result was his Model T design of 1908. Ford's "Tin Lizzie" was built in a factory using new conveyor belts and production-line techniques. Ford sold 15 million Model T cars in the years 1908 to 1927 and became one of the world's richest industrialists.

Karl Benz showed little interest in these new developments. He distrusted any mechanical part he had not watched being made by a worker at a bench. By 1900, Benz cars had hardly changed from those of ten years earlier, although they were reliable.

Julius Ganss, the sales director, watched Benz sales slump and became anxious. He ignored Benz's furious opposition and imported a shaft-driven Renault. He hired French technicians to copy it and build a modern, shaft-driven Benz. The result was not a success. This was damaging for Benz, as the new Mercedes cars, built by Maybach, were very successful. The company went through a crisis in 1903 and 1904. Benz who had been concentrating on a new four-cylinder engine, resigned briefly from the company.

Benz eventually agreed to allow his cars to race, and racing success eventually rescued the company. By 1908, Benz had a 120-hp racer, a monster that roared from St. Petersburg (now Leningrad) to Moscow in

Henry Ford's 1908 Model T was the first car to be mass-produced. For millions of people, the "Tin Lizzie" provided their first experience of motoring.

eight-and-a-half hours. An average speed of 50 mph (80 kph) over Russian cart tracks was an incredible feat. In 1909, the 200-hp Blitzen Benz was driven for the first time at Brussels in Belgium. It covered 1 mile (1.6 km) in just over 31 seconds.

Benz cars now won fresh admirers. The *Autocar* magazine tested a four-seater touring car in 1911 over roads that were "in a condition for which those responsible should be thoroughly ashamed." The magazine was impressed; potholes were scarcely noticeable, gear changes were smooth, both brakes were effective, the steering light, and the acceleration, even on hills, was good without any laboring.

By 1914, cars, trucks and buses were found on the roads in growing numbers. In 1910, some "village women of the United Kingdom" sent a petition to Queen Mary, complaining about the automobile. "Our children are always in danger, our things are ruined by dust, our rest is spoiled by the noise at night."

After World War I (1914–1918), Germany was in political and financial ruin. In 1926, the Benz and Mercedes (Daimler) companies merged to form Daimler-Benz. One of the company's bright young engineers was Ferdinand Porsche, whose name was also to become famous in the car industry.

Above *The Blitzen Benz racing car, designed by Hans Nibel, was capable of more than 125 mph (200 kph)*

Below *A modern Porsche 911. Ferdinand Porsche (1875-1951) learned his design skills working in the Daimler-Benz factory.*

The car industry

It was the invention of the internal-combustion engine that hastened the development of the automobile. Karl Benz's three-wheeler, built in 1885, was probably the first car of this kind. By the turn of the century, cars had begun to catch on in Germany, France and the United States. (In Britain the Red Flag Act, 1836–1896, made cars impractical.) By the 1890s the "horseless buggy," with its internal combustion engine, was being manufactured in the United States. Then, in 1907, the Ford Motor Company was formed. The conveyor belt and the technique of assembly line production made possible the first standardized, inexpensive car. Ford became the largest auto producer in the world, and Americans became the most devoted and numerous users of automobiles.

Industrial robots operate an assembly line at the Daimler-Benz factory.

Karl Benz made his last public appearance in 1925, riding in one of his first cars at a jubilee parade. He died on April 4, 1929. His wife Berta lived to be 95, dying in 1944.

Benz's first three-wheeler was primitive. Even some of his ideas in the more sophisticated four-wheelers were abandoned in the early 1900s, as engineers came up with new solutions for the developing problems of car design. For instance, Benz persisted with rear-engined drive after his main rivals had switched to front-engined cars. However, his innovations created a pattern against which others measured their own ideas.

Benz lived to see the automobile become very popular. Today, in many countries, most people own cars and could not imagine life without them. As a result, motor vehicles clog city roads, jam highways, pollute the air and cause many thousands of deaths each year through traffic accidents. Because of this, perhaps the gasoline engine will not power cars in the next century; another type of engine will have to be developed. However, Karl Benz will always be remembered for his contribution to a technology that has changed so many people's lives.

Traffic on a major highway in England. In Karl Benz's lifetime, the automobile became a universal phenomenon. Today, in most developed countries, so many cars crowd the roads that restrictions on car use may not be too far in the future.

Date Chart

1844 Karl Benz born at Karlsruhe, Germany.
1861 Studied gas engine in Stuttgart factory.
1864 Graduated from polytechnic. Worked first as a locksmith, then as a factory fitter.
1870 Death of his mother, Josephine Benz. Worked at Pforzheim as an engineer.
1872 Married Berta Ringer. Started own business.
1878 At work on a two-stroke stationary engine, finished by early 1881.
1883 Developed electric ignition system.
1885 Tested three-wheeled car with a four-stroke gasoline engine, some months before Gottlieb Daimler's gasoline-engined bicycle.
1886 Car made first public runs in Mannheim.
1888 Berta Benz and her two sons drove the car from Mannheim to Pforzheim.
1893 Benz's first four-wheeled car, the *"Vis-a-Vis,"* with passengers facing the driver. Benz "Victoria" and "Velo" models in production, and widely sold.
1895 Largest Benz car to date is the 8-seater "Phaeton."
1897 Benz's *"Dos-a-Dos"* (Back-to-Back) model appeared, with a 9-hp engine and a top speed of 25 mph (40 kph).
1901 Ransom Eli Olds began mass production of cars in the United States. Benz sales slumped because of competition from more up-to-date cars.
1909 Blitzen Benz car, designed by the firm's chief engineer Hans Nibel, set world speed records. Racing success helped the Benz company to recover.
1914–18 World War I.
1923 Benz began selling diesel-engined trucks and tractors.
1924 Benz and Mercedes (Daimler) companies agreed on merger, which took place two years later.
1925 Benz's last public appearance, at a Munich auto jubilee parade.
1929 Karl Benz died at his home in Ladensburg.

Books to Read

Automobiles by Lu Keatly (Specialty Books Intl, 1979)
The Automobile: Inventions That Changed Our Lives by Barbara Ford (Walker & Co., 1987)
Henry Ford: Automotive Pioneer by Elizabeth Montgomery (Garrad, 1969)
Henry Ford: Boy with Ideas by Hazel B. Aird (Macmillam, 1959)
Super Car by Mike Trier (Watts, 1988)

Picture acknowledgments

Mary Evans 4, 5 (top), 6 (lower), 8 (top), 9, 10, 12, 13, 16, 18, 20, 40 (both), 41, 42; Hulton-Deutsch Collection *cover*; Ann Ronan 6 (top), 23, 24 (both), 32; Wayland Picture Library iii, 5 (lower)/ Mercedes-Benz, 7, 8 (lower), 11, 14, 15, 27, 28, 29, 30, 31, 33, 34 (top); Zefa 34 (lower), 36 (lower), 43 (lower), 44, 45; Artwork on pages 17, 19, 21, 22 and 25 is by Jenny Hughes. Cover artwork is by Richard Hook.

Glossary

Accumulator A type of battery for storing electricity, used to provide the ignition power source.

Camshaft A shaft with oval lobes (cams). As the shaft spins, the cams push open the engine valves.

Carburetor The part of an engine in which gasoline and air are mixed to form an explosive mixture.

Combustion The process of burning; in the internal combustion engine, gasoline is burned in air.

Compression Squeezing; the gas-air mixture inside the cylinder is squeezed into a smaller space by the rising piston.

Crankshaft A shaft linked to the rod on the end of the piston. The crankshaft changes the up-and-down-motion to a round-and-round motion.

Cylinder A tube-like chamber inside which the piston moves up and down.

Diesel engine An internal combustion engine invented by Rudolf Diesel of Germany in 1892. Fuel is ignited by hot compressed air, not by spark plugs.

Expansion Spreading out, taking up more space. Materials usually expand when heated.

Flywheel A spinning wheel attached to the crankshaft. It stores enough energy during the power stroke to carry the crankshaft through the other three strokes.

Four-stroke engine An engine in which the piston makes four movements, or strokes, to produce power.

Friction The force created when two substances rub together.

Gears Wheels with teeth, or cogs, that engage, or interlock, with each other.

Horsepower (hp) A unit of power; 1 hp equals 745.7 watts.

Ignition system An electrical circuit that uses power from a battery to produce a spark to explode the fuel mixture inside the cylinder.

Induction In a four-stroke engine, this means the sucking in of gasoline and air as the piston descends.

Patent A document that protects an invention from being copied.

Piston A drum-shaped part that slides up and down snugly inside the cylinder. It is linked by a rod to the crankshaft.

Pneumatic Filled with air.

Pulley A grooved wheel, with a belt, chain or rope running around it, used in factory machinery and cars.

Spark plug A device in which a spark jumps across a gap when a strong electric current passes through it.

Steam engine An engine in which pistons are driven by the force of high-pressure steam.

Stroke The movement of the piston inside the cylinder.

Suspension In early cars, metal springs that absorbed jolts on bumpy roads.

Transmission The system that links the engine to the drive wheels.

Two-stroke engine An engine in which power is produced for every two strokes of the piston.

Vacuum A space from which air (or any other gas) has been removed.

Valve A device that lets a substance flow in one direction only. For example, inlet valves open to let fuel into the cylinder. Exhaust valves open to let spent gas out.

Index